DECOMPOSERS

in the Food Chain

ALICE B. McGINTY
Photography by DWIGHT KUHN

The Rosen Publishing Group's
PowerKids Press™
New York

To my mother, Linda K. Blumenthal — Alice B. McGinty
To my mother and father — Dwight Kuhn

Published in 2002 by The Rosen Publishing Group, Inc.
29 East 21st Street, New York, NY 10010

First Edition

Book Design: Maria Melendez
Project Editor: Emily Raabe
Photo Credits: p. 14 © Digital Stock; p. 18, folios © Brian Kuhn; all other photos © Dwight Kuhn.

McGinty, Alice B.
Decomposers in the food chain / Alice B. McGinty.
p. cm. — (The library of food chains and food webs)
Includes bibliographical references (p.).
ISBN 0-8239-5757-8 (lib. bdg.)
1. Biodegradation—Juvenile literature. 2. Food chains (Ecology)—Juvenile literature. [1. Food chains (Ecology) 2. Biodegradation. 3. Ecology.] I. Title. II. Series.
QH530.5 .M44 2002
577'.16—dc21

2001000171

Manufactured in the United States of America

Contents

Food Chains and Webs

Plants provide food not only for rabbits, but for insects, deer, people, and many other animals. Each of these relationships forms a food chain. Because most animals eat many types of food, they belong to many food chains. When many food chains are linked, they form a food web.

A new plant sprouts from the soil. One day a rabbit munches on the plant's leaves. A food chain has begun. Every time an animal eats a plant or another animal, they form a food chain. Food chains begin with **producers**. Producers, such as plants, make their own food. The next link in a food chain is made by the **consumers.** Consumers cannot make their own food. They must eat plants or other animals. Consumers that eat plants are called herbivores. Rabbits are herbivores. Meat-eating consumers are called carnivores. Wolves are carnivores. Some animals, called scavengers, eat the bodies of dead plants and animals.

Scavengers are also consumers. Every food chain ends with **decomposers**. Decomposers break down the bodies of dead plants and animals. In this book, you will learn why decomposers are so important to the food chain.

Decomposers break down dead plants and animals, like the dead leaves shown here. Decomposers make it possible for new plants to grow and new food chains to begin.

Energy and Nutrients

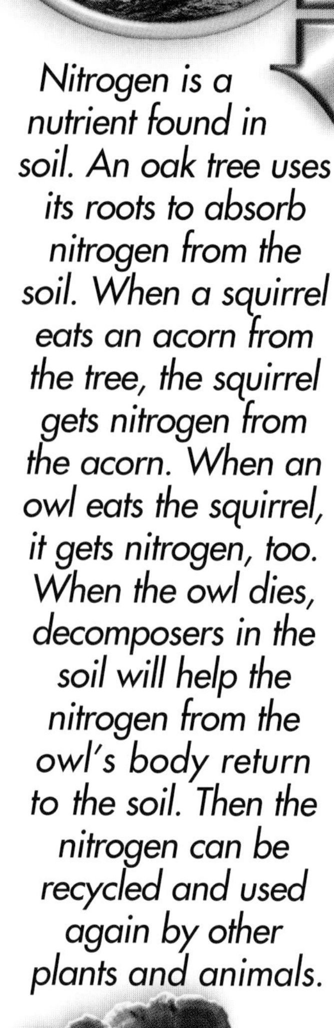

Nitrogen is a nutrient found in soil. An oak tree uses its roots to absorb nitrogen from the soil. When a squirrel eats an acorn from the tree, the squirrel gets nitrogen from the acorn. When an owl eats the squirrel, it gets nitrogen, too. When the owl dies, decomposers in the soil will help the nitrogen from the owl's body return to the soil. Then the nitrogen can be recycled and used again by other plants and animals.

Energy enters each food chain through sunlight. Producers use energy from sunlight to make food. Animals get energy by eating the producers, or one another. Energy is passed from link to link up the food chain. With each link, animals use up energy to grow and to move around. Luckily the Sun provides an unending supply of energy for food chains.

Plants and animals also need **nutrients** to live and grow. Plants get many nutrients from the soil by absorbing, or taking them in, through their roots. When an animal eats a plant, the animal gets nutrients from the plant. Nutrients are passed up the food chain from one link to the next. Unlike energy, nutrients have only a limited supply. Nutrients must be **recycled**, or used again and again. Decomposers return the nutrients to the soil so that plants can use them again.

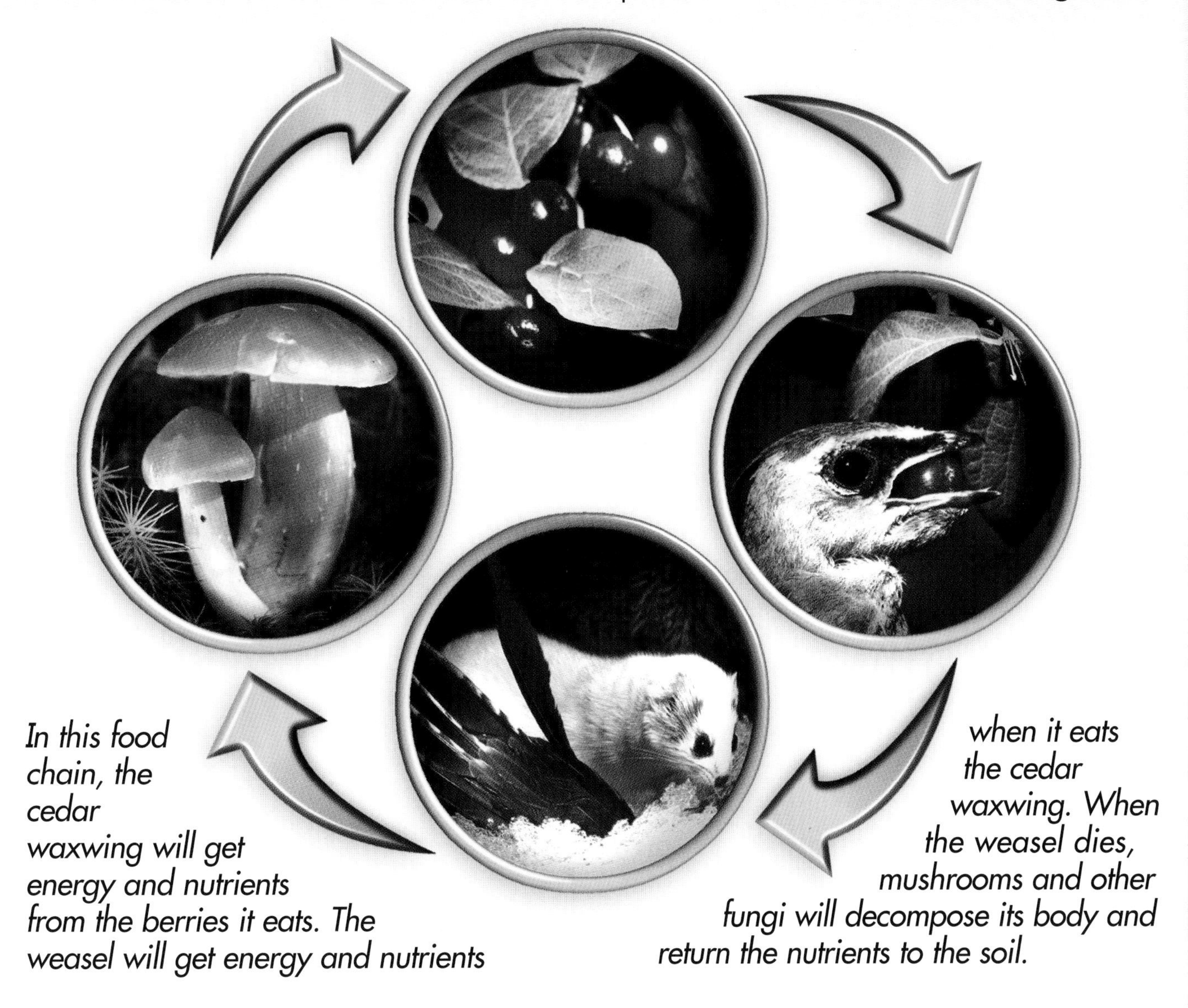

In this food chain, the cedar waxwing will get energy and nutrients from the berries it eats. The weasel will get energy and nutrients when it eats the cedar waxwing. When the weasel dies, mushrooms and other fungi will decompose its body and return the nutrients to the soil.

Decomposers on the Forest Floor

We do not see many dead bodies or waste materials on the ground. Where do they go? Decomposers break them down. If it weren't for decomposers, Earth would be piled high with dead things!

The decomposers in a forest help ensure that no nutrients are wasted. Forest animals, such as deer and birds, eat leaves, seeds, and fruits from trees. These animals eat only a small part of each tree. They do not eat the trunk, branches, or many leaves on the trees. These things fall to the ground when the trees die. Dead plants and animals are called **detritus**. The detritus on a forest floor may include tree trunks, leaves, animal droppings, and the bodies of dead animals. Detritus has many nutrients in it. Plants, however, cannot use the nutrients in detritus. Plants cannot absorb dead leaves and animals through their roots. The nutrients must be broken down and changed into a different form so that plants can use them. Decomposers break down the nutrients in detritus into a form that plants can use.

Decomposing logs contain a lot of water and nutrients. Sometimes new plants sprout inside them. As the logs decompose, they become part of the soil. The new plant will have rich soil surrounding its roots.

The Shredders

First in the lineup of decomposers are the animals that eat detritus. Beetles, earthworms, millipedes, termites, pill bugs, snails, and slugs all eat detritus. These creatures are called shredders. Shredders chew, or shred, dead wood and leaves into smaller, softer pieces. They break detritus down so smaller decomposers can eat it. Shredders **digest**, or break down, the detritus inside their bodies to form nutrients. They use some of the nutrients to grow. Extra nutrients pass through their droppings into the soil. Earthworms mix the detritus with their own droppings as they tunnel through the ground. This brings the detritus closer to the many small decomposers that live underground. The

The word millipede means "a thousand legs." Millipedes were given their name because they look like they have thousands of legs. Most millipedes actually have less than 300 legs. Millipedes have powerful jaws that help them chew up tough leaves that have just begun to decay.

earthworms' tunnels also help air and water move into the soil. Plants and the smaller decomposers need this air and water to live.

This earthworm is pulling a dead leaf under the ground to eat it. Earthworms also eat stems, roots, and small pieces of dead animals that are mixed into the soil. Shredders such as earthworms are sometimes called scavengers because they eat dead plants and animals.

Saprobes

Saprobes such as these penicillium *molds, shown here growing on oranges, are molds that are extremely useful to humans.* Penicillium *is used to make the antibiotic called* penicillin. *Penicillin is used to fight infections in people and animals.*

After shredders break the detritus into small pieces, other decomposers can go to work. What are these other decomposers? They are called **saprobes**. The word saprobe comes from a Greek word meaning rotten. Most saprobes are tiny organisms. They can be seen only through a **microscope**. Saprobes live in the soil on pieces of detritus. They eat the detritus by releasing chemicals called **enzymes** onto it. The enzymes break down the detritus. This releases the nutrients that are in the detritus. A saprobe absorbs some of these nutrients, while the rest are washed away into the soil. These extra nutrients then can be used by plants.

The mold on fruit is a kind of saprobe.

As the mold digests this peach, it breaks down the peach to release its nutrients.

The mold will keep growing and spreading until the peach is completely decomposed.

Bacteria

Bacteria are one kind of saprobe. Bacteria are tiny organisms that live almost everywhere. You can find bacteria in the air, in the soil, in water, and inside your body. Some kinds of bacteria are harmful. They are the germs that cause disease. Most bacteria are helpful. There are bacteria inside your body that help digest food. There are many helpful bacteria in the soil. These bacteria are decomposers. Bacteria in the soil feed on detritus. Like other saprobes, bacteria release enzymes onto their food to get nutrients. The bacteria use the nutrients to grow and to make more bacteria. The more detritus there is, the more bacteria will

Some bacteria live on the dark bottom of the sea. They feed on tiny pieces of dead plants and animals that have settled there. The bacteria feed on this detritus and release the nutrients inside it. The nutrients drift back to the surface in the ocean currents. There the ocean's producers can use the nutrients to grow and begin new food chains.

grow in the soil. One scientist has said that there may be 10 million bacteria in one spoonful of farm soil!

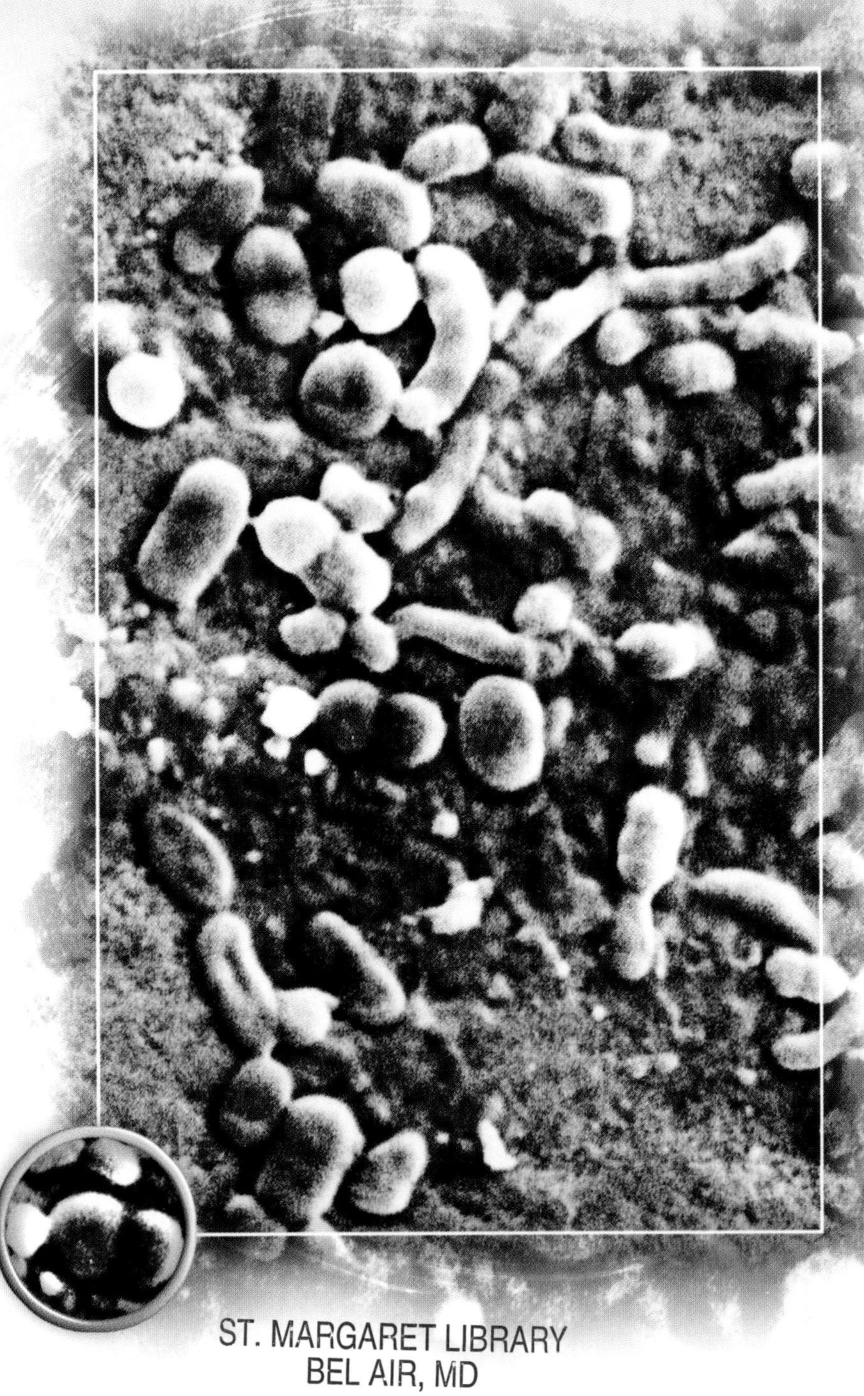

These bacteria might be living in your mouth right now! They are the bacteria found in plaque, a sticky film that grows on teeth.

Fungi

Fungi are another common kind of saprobe. Fungi are plantlike organisms. However, fungi are not green, do not make their own food, and do not need light to survive. Most fungi live underground, underwater, and in dark, damp places. Mushrooms, molds, and yeasts are types of fungi. You may have seen mushrooms growing in the ground or mold growing on rotting fruit. What you see, however, is only a small part of the fungus. The main part of the fungus lives under the ground or inside the rotting fruit. This main part of a fungus is made up of thin threads. Each thread is called a **hypha**. The fungus eats by releasing enzymes from each hypha. The

You can see mushrooms growing on dead logs, on lawns, and in places where dead leaves are on the ground. These are the places where a fungus has food. Some mushrooms can grow in many places because they eat many kinds of detritus. Other kinds of mushrooms eat only certain foods and can live only in certain areas.

enzymes digest the food, and the hyphae absorb nutrients from the food. Fungi also absorb water and air through their hyphae.

This mold is growing on a tomato. The mold will grow until the entire tomato is rotted away.

In this close-up photograph, you can see the mold's hyphae growing into the fruit.

How Fungi Spread

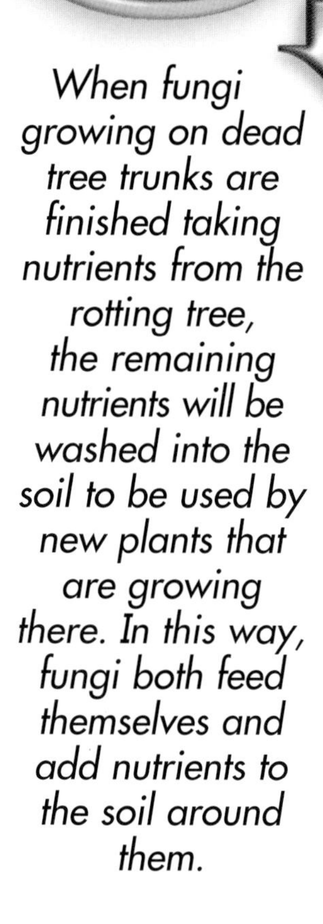

When fungi growing on dead tree trunks are finished taking nutrients from the rotting tree, the remaining nutrients will be washed into the soil to be used by new plants that are growing there. In this way, fungi both feed themselves and add nutrients to the soil around them.

The parts of fungi that we see, such as mushrooms or fuzzy mold, are called **fruiting bodies**. Fruiting bodies help the fungi **reproduce**, or make more fungi. Fungi reproduce by spreading **spores**. Spores are like tiny seeds. Each fruiting body produces millions of spores. The spores of a mushroom are underneath its umbrella-like top.

Fungi spread their spores in many ways. Some mushrooms drop their spores onto the ground. The spores are light and can be spread by wind or rainwater. Puffball fungi explode to scatter their spores. If a spore lands on a place with food, it will grow. The more detritus there is to land on, the more new fungi will grow. After its spores are released, the fruiting body usually dies. Later the fungus will make more fruiting bodies.

These are examples of the fruiting bodies of some molds. The molds are (clockwise from top left): *molds on a pumpkin, gray snow mold on an apple, molds on bread, and molds on a tomato.*

Slime Molds

Slime molds are interesting saprobes. They grow on damp, rotten wood and on decomposing leaves. In their creeping stage, slime molds move like oozing jelly. The slime mold surrounds its food and releases enzymes to digest the food. Slime molds feed on pieces of decaying leaves, rotten wood, parts of dead animals, and bacteria.

When the slime mold runs out of food or water, it changes from its creeping stage to its spore-forming stage. In its spore-forming stage, the slime mold grows large lumps, called spore cases, which hold spores. In time these spore cases will burst and release

Many kinds of slime molds are named after the foods they look like, such as pretzel slime, scrambled egg slime, and chocolate tube slime.

the spores. Many slime molds change their shape, color, and texture during each of their two stages. In only hours, a slime mold may change from being gooey to being soft and fluffy.

This slime mold is in its creeping stage. It is creeping along a piece of wood that was found lying in the soil.

This slime mold is changing into a fruiting body. The fruiting body will release spores when it is complete.

This slime mold is in its spore-forming stage. The pink objects are the slime mold's fruiting bodies.

A New Beginning

When people cut down trees in a forest and take them away to use as lumber, the decomposers cannot do their job, and the forest loses the nutrients that were in the lumber. If enough trees are taken away, the forest soil will not have enough nutrients for new trees to grow.

By returning nutrients to the soil, decomposers help new producers to grow. Without producers, there would be no life on Earth. Producers provide the nutrients that are passed to every member of the food chain. Decomposers make it possible for producers to live. You can help decomposers by starting a **compost heap**. Fill it with dead leaves, grass clippings, and leftover foods, such as vegetable peels. The decomposers will turn the dead plants and food waste into nutrients that make rich soil. Look closely at your compost heap to see the shredders and fungi at work. If you are lucky, maybe a new plant will sprout in the soil.

Glossary

bacteria (bak-TEER-ee-uh) Tiny living things that can be seen only with a microscope and can sometimes cause decomposition or illness.

compost heap (KAHM-post HEEP) A mixture of decomposing food and plant material.

consumers (kon-SOO-merz) Members of the food chain that eat other organisms.

decomposers (dee-kum-POH-zers) Organisms, such as fungi, that break down the bodies of dead plants and animals.

detritus (dee-TRY-tus) Dead plant and animal material.

digest (dy-JEST) To break down the food and use it for energy.

enzymes (EN-zymz) Chemicals that are made by plants or animals and cause changes in other substances.

fruiting bodies (FREWT-ing BAH-deez) The parts of fungi that produce spores for reproduction.

fungi (FUHN-jy) Non-green organisms that feed on living or dead organisms.

hypha (HY-fah) A thread-like tube that is part of a fungus.

microscope (MY-kroh-skohp) A piece of equipment that makes small things look larger.

nutrients (NOO-tree-intz) Anything that living things need for energy or to grow.

producers (pruh-DOO-serz) Plants and algae that use sunlight to make their own food.

recycled (ree-SY-kuhld) Being used again in a different way.

reproduce (ree-pruh-DOOS) To make more of something of the same kind.

saprobes (SAH-probz) Organisms that feed on dead plant or animal material and that digest their food outside their bodies.

spores (SPOHRZ) Tiny seedlike packages that are produced by fungi for reproduction.

Index

Web Sites

To learn more about decomposers in the food chain, check out these Web sites:

www.fortworthgov.org/SWMservices/let_it_rot/

www.nhptv.org/natureworks/nwep11.htm